# Conference on Engineering Cosmetics and Consumer Products (ECCP 2019)



Orlando, Florida, USA
9 – 10 November 2019

ISBN: 978-1-7138-0573-1

**Printed from e-media with permission by:**

Curran Associates, Inc.
57 Morehouse Lane
Red Hook, NY 12571

**Some format issues inherent in the e-media version may also appear in this print version.**



Printed with permission by Curran Associates, Inc. (2020)

For permission requests, please contact AIChE
at the address below.

AIChE
120 Wall Street, FL 23
New York, NY 10005-4020

Phone: (800) 242-4363
Fax: (203) 775-5177

www.aiche.org

**Additional copies of this publication are available from:**

Curran Associates, Inc.
57 Morehouse Lane
Red Hook, NY 12571 USA
Phone: 845-758-0400
Fax: 845-758-2633
Email: curran@proceedings.com
Web: www.proceedings.com

# TABLE OF CONTENTS

## PROCESS ENGINEERING



## PRODUCT ENGINEERING



## PACKAGE ENGINEERING



## OPEN INNOVATION

## REGULATORY ISSUES, SAFETY, COMPLIANCE



## SUSTAINABILITY



## POSTER SESSION

# Innovative Cosmetics powered by Emergent Technologies

Authors:
Bergmann, W., Unilever

Recent advances in major technology areas have been used in cosmetic innovations to address increasing consumer demands on how cosmetic materials deliver benefits and navigating a shifting regulatory environment. Examples of these technology areas include microbiome, sustainable materials (packaging), protection against environmental stressors, and cosmetic benefits sourced from nature. The presentation will share examples of how these technologies have been used in cosmetic products launched over the last 12 months.

**Effective Design of Particulate Products and Processes - a Systems Perspective**

Authors:

Mort, P. R., Procter & Gamble co.

Product quality and processing efficiency are both paramount in consumer products industries. Model-based approaches can improve the design of particulate products and lead to more efficient processing. This seminar will discuss a complement of modeling and experimental efforts that are relevant to consumer-based particulate products, specifically in the context of the design and execution of Purdueâ€™s Center for Particulate Products and Processes, CP3, including:

- Fundamental research in powder flow and stress fields, a foundational element of processing.
- Intensified formulation: granulation with highly concentrated [active] binders.
- A modeling framework: formulation, control, scale-up, and systems integration.
- Reconsidering flowsheet modeling: how to facilitate its adoption?
- Modular solids process design: an enabler for process intensification and optimization.

Further, we will discuss innovation as a collaborative strategy. Innovation is necessary to fuel growth in industry; yet bold innovation can be risky especially when core inventions rely on having particles in product design and/or processing operations. How can we become more effective in particulate product innovation? We often see correlation between successful innovation and useful models. While models are often used to quantify fundamental aspects of an invention, they can also be used to refine a problem statement, define more robust and efficient means of validation, and even to open up new ideas for execution and acceptance of an innovative product or process in the market. Overall, modeling helps us to deepen our understanding, extend breadth of connections, and can be used as a basis for collaboration within organizations as well as external partnerships.

**Tunable Transport Kinetics and Mechanics of Organohydrogel Fibers**

ECCP Engineering Cosmetics & Consumer Products

---

Authors:

Hsiao, L., North Carolina State University

There has been a surge of interest in composite hydrogel fibers with tunable transport properties, in addition to advantages such as increased moisture holding, mechanical enhancement, and biocompatibility. A major challenge is that it is difficult to generate fibers with functional microstructure without expensive post-processing steps. We introduce a colloidal gel system comprised of nanoemulsion droplets of poly(dimethylsiloxane) suspended in a continuous phase comprised of a double network hydrogel precursor. The nanoemulsions undergo self-assembly at elevated temperatures to form hydrophobic networks with different length scales, which can be crosslinked by UV light due to its hydrogel phase. We find that the length scales of the colloidal nanoemulsions are tuned independently using the applied shear rate and temperature in a cylindrical spinneret, generating solid organohydrogel fibers with different microstructures. Furthermore, we show that it is possible to introduce a double network hydrogel precursor that is compatible with the self-assembling nanoemulsions, thus generating fibers with enhanced Young's modulus, fracture energy, and rupture stress. This study utilizes state-of-the-art understanding of colloidal gel rheology and hydrogel chemistry to generate new types of mechanically reinforced materials that can also be used to control drug loading and release.

## Green Extraction with a New Lipotropic Solvent for Performant Biomimetic Skin Actives: OSMOSTMconcept

Authors:
Sibra-Lemoine, N., Hallstar

The unique and high-performance green OlÃ©o-Ã©co-extraction technology, developed and patented in France, is built on the idea of gathering the best from nature to cover a broad range of functions. Through this technology, ingredients of high purity and yield are produced in an ecologically sustainable way, without the use of organic solvents and with no risk of pollutants.

The resulting actives â€“ called OlÃ©oactifÂ® â€“ are holistically developed with a combination of several plants and oil-based solvents. OlÃ©oactifÂ® are created using vegetable oil to transfer the biological footprint of a plant directly into an oil-based complex. They are highly concentrated biomimetic actives that are easily formulated into virtually all cosmetic products.

Oléo-eco-extraction (OEE) is an eco-process intensified by high energy microwaves (MW) and low frequency ultrasounds (US) using natural triglyceridic oils as green, safe and bioactive extractants. To reduce production time and energy, a new co-intensified [MW+US] process has been applied with an innovative biomimetic solvent inspired both from plant membranes and skin cells. This concept, called OSMOS$^{TM}$, is able to increase the extraction yield of phyto-molecules as well as their bioavailability as a skin delivery system.

The OSMOS$^{TM}$ concept provides a new plant and skin biomimetic system for green intensified eco-extraction of cosmetic actives that are safe, non-oxidized, stable, dry-touch, organic certified and highly bioavailable.

## Influence of Different Process Variables in the Final Characteristics of an Oil in Water Cosmetic Emulsion: A Scale-up Strategy

Authors:
Suaza, A., Universidad Nacional de Colombia
Orjuela, A., Universidad Nacional de Colombia
Rodriguez, H., Universidad Nacional de Colombia

Global cosmetics market has been steadily growing during last years. A major driver of this, is the consumer need for innovative products with multifunctional attributes. As the market is highly competitive, there is need to reduce the launch time of products, thus becoming in a dominant constrain during product design.

Despite different tools have been developed for *in-silico* product conception and design, a limiting factor still remains in the scale-up process. In particular, when developing structured products in the cosmetics industry, it is necessary to consider several variables to ensure that the product will fulfill the desired properties and attributes, but also will maintain a long-term stability. Variables such as stirring rates, equipment geometry, emulsification temperature, mixing times, heating or cooling rates, and ingredients addition procedures play a major impact in the final characteristics of the emulsified products. Some of these variables need to be adjusted to account for the change in the mixing-volume-to-transfer-area ratio. Currently, the scale-up to industrial production is a time-consuming, resources-intensive process, traditionally involving trial and error techniques.

In this regard, different systematic scale-up techniques have been developed for the production of emulsion. Specifically, the use of process invariants such as dimensionless numbers of the form $ND^{Î±}$, have been integrated as key parameters during change of scale. While these techniques mainly focus in maintaining a defined drop size distribution, it has been verified that this doesn't guarantee to maintain all the characteristics of the emulsion.

Thus, the aim of this work is to study the influence of some process variables, focusing in the change on the heat transfer conditions during scale-up, and their impact in the characteristics of an oil in water emulsion. The study was carried by the manufacture of a cosmetic emulsion at three different processing scales in order to identify suitable scale-up strategies.

**Digital Thread for Design, Engineering & Manufacturing for Consumer and Cosmetics Industries**

---

Authors:

Aglave, R., Siemens PLM Software

Kopcha, S., Siemens

Picciotto, D., Siemens

Bouvet, C., Siemens

Longwell, G., Siemens

Consumer industry faces many challenges: Performance analysis of 34 of the world's top 50 consumer goods companies shows that 85% of those big companies had seen a decline in either revenues, profits or both. Only 15% managed to escape unscathed. Due to the new health and environmental awareness of consumers, companies need fast time to market to responds to their needs. There is also a trend of smaller batch sizes to cater to diversified and to some degree individualized needs

This means that consumer goods companies are innovating with new recipes and ingredients. Changing recipes of ingredients many times requires a revamp or reestablishment of engineering practices.

Smart process design uses digital representation of the manufacturing unit operations to evaluate performance of those steps. Starting from advanced engineering simulation to create a performance twin of the process, one can perform many trials for improved design to meet process requirements and optimize the operations. This performance twin then helps in tearing down the walls to the manufacturing process itself, resulting in an operation twin. Through automation, control simulation and virtual commissioning, the transformed recipes can be handed over to manufacturing sites rapidly.

In this presentation, we will describe what such a transformation means as well as the tools that are necessary to carry those out. We will also present a few case studies where it had been successfully utilized.

**Multiple Emulsions: Scaling up and Scaling Down for Formulation Design**

---

Authors:

Helgeson, M. E., University of California, Santa Barbara

Multiple emulsions (i.e. droplets within droplets) are attractive encapsulation and delivery vehicles due to their ability to co-encapsulate compounds with widely varying solubility within their multi-phase structure. As such, they have been explored for a range of formulated fluid technologies including consumer products, foods, cosmetics, pharmaceuticals and nanobiotechnology. Recent advances in microfluidic methods to produce multiple emulsions with exquisite control have opened up a wide new design space of complex multi-phase droplets, and various colloidal particles from them. However, these capabilities have yet to meet the demands of technological translation due to problems with scalability in two important respects. First, the devices used have yet to demonstrate production throughput at levels necessary for industrial implementation. Second, the size of droplets is set by the device geometry, which has limited production to droplets that are tens or hundreds of microns in size too large to achieve colloidal stability in many conventional formulations. Here, we review several recently developed methods to produce multiple nanoemulsions (i.e. nanodroplets within nanodroplets) that overcome these limitations in scalability. By combining conventional, scalable emulsification equipment with careful control of mechanical and thermodynamic instability within droplets and at their surface, these methods are able to produce large quantities of colloidally stable multiple emulsions with well-specified morphologies. These complex droplet morphologies are retained upon the addition of various material pre-cursors, and the droplets are stable over the time scales required for material chemistry, thereby enabling their use as templates for complex nanoparticles. As a detailed demonstration, we show how oil-in-water-in-oil double nanoemulsions can be used to template the synthesis of oil-filled nanogels for encapsulation and controlled delivery of poorly water-soluble active ingredients.

# Amphiphilic Polymer Structure Confers Function for Formulation Engineering

Authors:

Alexandridis, P., University at Buffalo, The State University of New York (SUNY)

Tsianou, M., University at Buffalo, The State University of New York (SUNY)

Amphiphilic polymers, encompassing synthetic block and graft copolymers and polymers of biological origin such as polysaccharides and proteins, exhibit an innate ability to organize from the nanoscale across to the mesoscale, in the bulk and on surfaces. Such structuring leads to function, and thus amphiphilic polymers find applications in diverse formulated products. The solvent, typically water, and the additives present therein, e.g., glycols, salts, particles, provide valuable degrees of freedom for controlling the morphology and, hence, structure/property relationships. Furthermore, solvents and particles or surfaces can dramatically affect the molecular mobility and the dynamics of structural transformations.

The presentation will utilize principles of polymer organization in solution and on surfaces to highlight examples of (i) how polymer solution association and surface adsorption can be modulated by the polymer chemistry, solvent additives, and/or surface properties, and (ii) how the resulting structure can lead to beneficial function for formulations. Amphiphilic polymers of the poly(ethylene oxide)-poly(propylene oxide) type and those resulting from polymer-surfactant complexation will be used as case studies. The polymer self-assembled structure in the bulk solution will be related to that on surfaces.

**Innovative Strategies for Broadening the Formulators Toolbox**

Authors:

Potuck, A., GlaxoSmithKline

Topical gel formulation enables delivery of therapeutics and active compounds not amenable to formulation as traditional biphasic creams. Gel formulations are often manufactured under ambient conditions and can contain non-aqueous solvents to mitigate potential for API (active pharmaceutical ingredient) degradation via thermally or hydrolytically labile actives. Driving innovation within the field of gel formulation and characterization promotes a deeper process and product understanding, ultimately delivering a more robust product to the patient. Given the trend of poor solubility of small molecule therapeutics, solvent systems with low or no water are increasingly demanded to boost solubility. This presentation will touch upon two case studies that showcase advancement in gel product development for poorly water-soluble APIs: 1) Solvation monitoring in non-aqueous gels, and 2) Critical material attribute identification of neutralizers in low water content gels.

# Composite Liquid Products: Design and Performance

Authors:

Spicer, P. T., UNSW Sydney

Caggioni, M., Procter and Gamble Co.

Squires, T. M., University of California, Santa Barbara

The demands made of formulated liquid products by consumers continue to increase in complexity. As a result, we require a more sophisticated ability to design novel functions, control consumer benefits, and manufacture these products robustly. Fluid products are still often conceived, formulated, and tested using assumptions of homogeneous behavior, or by focusing on microstructural elements for specific insights. However, new product design paradigms emphasize the use of multifunctional and hierarchical elements of fluid products, essentially as liquid composites, to deliver entirely new consumer benefits. Two fascinating aspects of complex fluid transport, inhomogeneous flow and two-fluid microstructures, will be demonstrated experimentally in formulated systems as a means of integrating such benefits into products. Novel rheometry techniques will be used to describe composite flows in formulated sparse networks of cellulose nanofibers. Once discussed as barriers to accurate bulk fluid characterization, we argue that these novel flows are now an enabler of new product features and increased efficiency and performance.

**Cellulose Derivatives As Rheology Modifiers for Cosmetic Applications**

ECCP
Engineering
Cosmetics
& Consumer
Products

---

Authors:

Sharma, V., University of Illinois at Chicago

Dinic, J., University of Chicago

Jimenez, L. N., University of Illinois at Chicago

Martinez, C., University of Illinois at Chicago

Cellulose derivatives are often used in cosmetics industry as rheology modifiers for formulating a large variety of consumer products. Conventionally, the industrial practice is to quantify shear rheology response over a very wide range of shear rates. A well-defined shear thinning behavior is known to provide formulations with high viscosity and additional stability against flocculation or precipitation during storage, and adequate viscosity reduction facilitates mixing and pouring as well as dispensing and drop formation. Even though substantial efforts have focused on the characterization of the shear rheology response of polysaccharide solutions, correlating material properties to processability in operations that involve drop formation and liquid transfer has remained a significant challenge. The missing, critical link in the processability puzzle arises due to unresolved challenges in characterization the macromolecular response to extensional flow, for shear-free extensional flow field gets created spontaneously in freely-draining liquid neck encountered during drop formation and dispensing operations. Using dripping-onto-substrate (DoS) rheometry protocols we developed recently, we characterize the pinch-off dynamics, and extensional rheology of aqueous solutions several cellulose derivatives to elucidate the influence of flexibility, extensibility and charge on the flow behavior and processability.

# Predicting Multiphasic Chemical Properties with Machine Learning

Authors:
Luxon, A., Virginia Commonwealth University
Ferri, J. K., Virginia Commonwealth University
McQuade, T., Virginia Commonwealth University
Le, Q., Virginia Commonwealth University
Parish, C., University of Richmond

Although chemistryâ€™s governing physics are largely understood, the number of features that determine the properties or performance of a product or formulation are too complex to wrangle for a real system of interest, such as an emulsion. This complexity is multiplied by intrinsic molecular features and extrinsic processing parameters both of which contribute to product performance. To model the effect of intrinsic features such as molecular structure or composition, engineers make reducing assumptions and apply mechanistic relationships to provide insight into aspects of formulations design. These models are often limited in scope and resolution. Machine learning (ML) provides the ability to tackle the high dimensional complexity of chemical systems but needs large data sets of physically relevant descriptions of molecules to create a useful model. A quantum mechanical (QM) wave function describes a molecule at the most fundamental level. A high-resolution approximation of the wave function can be calculated using density functional theory (DFT). We performed calculations on 4200 molecules for which experimental octanol-water partition coefficient (logP) values exist. The DFT geometry optimizations were carried out in the gas, water, 1-octanol phases using the Q-Chem software package. Quantum chemical descriptions of each molecule were extracted from the resulting wave functions. Supervised ML models were trained using logP as the target response and the QM descriptors as the model features. Multiple ML algorithms were used and evaluated based on prediction accuracy criteria. The resulting models have higher prediction accuracy and lower error than the current models using other molecular representations. It was found that polar surface area and polarizability are the most influential molecular features for predicting logP. Because DFT is relatively time consuming, the tradeoff between model accuracy and calculation time was investigated to elucidate the relationship between intrinsic descriptor resolution and predictive capacity in formulations consisting of multiphase systems.

**Holistic Product Design with L'Oreal's Packaging Innovation Team**

Authors:

Bickford, B., L'Oreal

The brands of ™Oreal are known for providing beautiful and captivating products that are built on not only high-performing formulations, but are caringly enclosed in creative packaging and applicators that are as eloquent as they are soundly engineered.

To create this holistic approach, elevating packaging and design to sit equally at the table with formulations provided by our outstanding research and innovation groups (R&I) is a challenging task. Packaging design does not start with pen to paper sketching or the rapid protoyping technologies that we have all seen glamorized on the surface of product development. The role of the designer starts far earlier with research and information gathering: we deeply understand our product and brands; we understand our global consumer's motivations, behavior and gestures; we define packaging parameters, select materials, and align with our formulation, production and manufacturing teams as close to the beginning of the project as possible. Even with all of this organization and information in place, we realize that there is still a great deal of abstraction and ambiguity in bridging these foundations to a 'must have!' product. Through packaging design we synthesize the brand desires to the brief, providing spontaneous and unconventional inventions and creative ideas at first as metaphors to understand what our marketing and creative teams like and dislike (both equally useful paths of information for design!), and iterate/tune these ideas using consumer-centric methodologies and feedback to quickly hone in on unique concepts that resonate and compel.

The precise mix of packaging design, engineering and formulation is as deeply the DNA of our stand-out products as are our individual brand attitudes and passion. The take away from this presentation is that 'to design' is not just creativity and beauty on its tangible face. Holistic packaging design elevates and differentiates our world class brands as a journey through rich user-experiences, learning and diverse story-telling; the bridging of technical complexities into human context; and putting safety and security at the forefront of all that we provide.

# Building Open Innovation Ecosystems: Trends, Technologies and Traps

ECCP
Engineering Cosmetics & Consumer Products

Authors:

Miller, B., yet2

Cosmetic and CPG companies have entered into an Open Innovation 3.0 landscape. They have moved far beyond basic initiatives to foster innovation in-house. They recognize that simply opening their doors and passively accepting ideas from individuals, customers and partners is both inefficient and insufficient to achieve innovation growth goals.

Today's industry leaders are actively putting in place the internal systems, processes and people to identify meaningful innovation opportunities and disruptive solutions and to capture and nurture them. No small feat given that traditional large companies operate entirely differently than the entrepreneurs and startups playing an increasingly larger role in innovation ecosystems.

Attendees of this 30-minute presentation will hear:

- Examples of how cosmetic and consumer products companies have used open innovation initiatives to identify game-changing product, engineering/manufacturing, logistics, and marketing solutions
- The four elements of a successful open innovation system
- Strategies for better interactions with entrepreneurs and startups
- Agreements, transitions, and adaptations large companies need to make to get to a 'deal' mindset

**Open Innovation in Cosmetics & Personal Care - Preparing for a New Future in Cpg**

ECCP
Engineering
Cosmetics
& Consumer
Products

---

Authors:

McLellan, K., The Clorox Company

Talk will cover past, present and future of what the Clorox Company has done to grow the Burt's Bees business through External, or Open Innovation. Also included will be an assessment of current best practices, and candid comments about what works best in the trends-driven world of Personal Care. Not all programs work out as intended, so talk will also highlight lessons learned and adjustments made along the way.

Established CPG companies have never been under more pressure to grow despite new external forces that cannot be managed based on the previous ways of working. How can Open Innovation identify and help build the new set of capabilities that enable businesses to successfully compete in the future?

The strategies used by Open Innovation have been evolving as well. This talk will discuss the advantages and potential pitfalls associated with traditional and new approaches including Technical Partnering, Expert Networks, Entrepreneurial Ecosystems and Data Analytics.

Talk will also address the latest thinking about how to measure impact and success, which can be elusive for practitioners, stakeholders and others in the business benefiting from Open Innovation's work.

Finally, the talk will wrap up with ideas about the importance of understanding an organization's culture and true appetite for change. Innovative cultures are often messy, and Open Innovation can help R&D leaders better prepare their teams for what comes next.

# Beyond Structure-Property Relationships: A Correlating Customer Reviews to Consumer Product Formulations

Authors:

Kay, K., Virginia Commonwealth University

Luxon, A., Virginia Commonwealth University

McQuade, T., Virginia Commonwealth University

Ferri, J. K., Virginia Commonwealth University

Multi-billion dollar industries such as pharmaceuticals, food, and consumer products rely on formulations with complex performance criteria. Relating the features of the formulation to consumer perception of the product is a significant challenge. This challenge arises, because both intrinsic and extrinsic physical and chemical descriptors, as well as personal and social factors affect public perception. User ratings and individual customer feedback have been used in many social platforms to improve and personalize user experience. We hypothesize that this feedback can also be used to guide manufacturers as well. To achieve this, a workflow relating ingredients and customer satisfaction must be established. We have developed a process and tools for ingesting, curating, modeling, analyzing and visualizing consumer review data from very large online retail distributors. Our hypothesis is that the intrinsic aspects of the formulated product, i.e. composition, and product performance - both of which are available from the retailer can be used to elucidate which if any correlations between the two exist.

Customer review data - including the list of ingredients for a wide range of consumer products are accessible online. This large array of feedback and product performance defines a hyperdimensional space of public perception of the product. Similarly, the product list also provides enough information to generate a list of CAS numbers. With some restrictions, CAS numbers can be used to generate molecular representations. From the CAS, a SMILES can be matched and used to make fundamental calculations of the molecular features associated with the product. These arrays of molecular descriptors are the molecular structure space. We present a path for relating the molecular and performance spaces of a wide range of retail consumer products sold online in order to enable advanced regression tools.

# Diffusing Wave Spectroscopy: Decoupling Physical Phenomena in Concentrated Systems

ECCP
Engineering
Cosmetics
& Consumer
Products

Authors:

McMillin, R. III, Virginia Commonwealth University

Ferri, J. K., Virginia Commonwealth University

Emulsions are the formulated form of a wide range of personal care and consumer products. A mechanistic understanding of the stability of these materials depends on the ability to decouple destabilizing effects, such as evolution of mean droplet size, the mean settling/creaming velocity of the dispersed phase, and the microrheology of the matrix phase. Typically, these effects are estimated using spatial scanning and average intensity of transmitted light through the sample. Although this approach can be used to identify when unknown dynamics have slowed or stopped, mechanical movement is required, inhibiting fast sampling times. Diffusing wave spectroscopy (DWS), a multiple scattering adjunct to photon correlation spectroscopy, has long been used as a tool to study concentrated and turbid samples. Due to its non-invasive nature, it represents an important technique to extract the structure and dynamics of sensitive and transient systems. DWS measurements are made on concentrated suspensions and emulsions in forward and back-scattering modes. Measurements made over the lifetime of complex formulations allows for accurate determination of stability and aging mechanisms. DWS presents a method of understanding what dynamics are dominant and how they are affecting the stability of multiphase systems. Further, utilizing light decorrelation in DWS provides near instantaneous measurements to be made, gaining access to extremely small sampling times. We show the utilization of DWS to provide quantitative determination of particle radius and settling velocity of polystyrene and SiO2 suspensions. We then use a similar framework to explore the role of surfactant concentration on the evolution of creaming velocity and average droplet radius in short-time aging in emulsions of isopropyl myristate and aqueous surfactant solutions.

# How to Market Personal Care Products Creatively While Being Compliant

Authors:
Aral, B., Hearst Corporation

An overview of federal as well as some notable state regulations will be provided. Key consumer watchdogs, industry groups and scientific organizations and their oversight will be outlined. Examples of non-compliant product claims or positioning will be given, to help avert legal action.

# Preservatives: The Ever-Changing World

Authors:

Pungitore, S., Symrise

Preservatives are an integral part of cosmetic formulations and consumer safety. They help to prevent microbial contamination, not only to protect the freshness of the product, but to also keep consumers safe from harmful microorganisms. In a world of changing consumer expectations, market trends, individual company policies and updated regulations, formulators need to be increasingly creative by using innovative antimicrobial ingredients and combinations to deliver safe and effective products to the personal care and cosmetic market. Let's review the current regulatory environment and toolbox of noncontroversial preservative and multifunctional ingredients available to industry professionals.

**Finding Health & Safety Buried Treasure with AI**

Authors:

Kimbleton, S., IBM

Garvin, T., IBM

The challenge to glean understanding and insight from an array of historical safety-related reports and observations has existed since the dawn of the HSE discipline. While most organizations today use traditional methods to analyze past events and activities along structured elements (time, place, risk rating and so on), a vast amount of wisdom around hazard identification, root causes and risk control measures remains buried in textual descriptions and reports, and teachable moments become lessons lost.

The hands and minds that developed these textual artifacts may be among the most seasoned in the organization, bringing years of experience to bear on the issues and opportunities involved. Such artifacts are then clearly buried treasure. Exploring and surfacing the insights contained in artifact repositories calls for new tools. Using these, a new type of H&S performance indicator could emerge: latent indicators, lying concealed within the written record, offering as much or more value as the leading and lagging indicators used today.

This paper describes leveraging the power of artificial intelligence (AI) to absorb large amounts of safety-related textual information, find common themes and identify similar events, which are then analyzed for patterns in causes and controls. This solution, used in concert with traditional analytics, offers unprecedented power to comprehend and visualize collective safety knowledge from historical record. Transforming words to wisdom in this manner not only illuminates the past but also provides a basis for actioning improvements in operational excellence.

# Biopolymer Hydrogels: Modulating Microstructure through External and Internal Stimuli

ECCP
Engineering
Cosmetics
& Consumer
Products

Authors:

Khan, S. A., North Carolina State University

Polysaccharide-based hydrogels are being sought in a range of application because of the wide array of properties that they offer. Of particular interest to us is the ability to tune their microstructure and rheology using external and internal stimuli. We examine three different polysaccharides, guar gum, aliginate and nanocellulose, to illustrate this notion. Guar gum with its mannose backbone and galactose side unit offers a unique opportunity to tailor its characteristics using back-bone of side-chain cleaving enzymes. While backbone hydrolysis leads to a loss in the entanglement, and ensuing network structure, we find that cleaving side-chains using enzymes lead to "hyper-entanglement" between guar polymer chains. This results in creation of gels, the structure of which is dependent on the extent of enzyme action and time. Nanocellulose, on the other hand, intrinsically forms a hydrogel in an aqueous media. However, exposure to different shear histories can lead to hierarchical breakdown of the microstructure, and reformation, leading to gels with different internal structure and rheology. We explore such restructuring the gel microstructure and its relation to modulus and yield stress Finally, alginate hydrogels respond well to UV light. Using calcium carbonate nanoparticles, and UV induced photo acid generators, physical crosslinks can be created to form alginate hydrogels with a variety of elastic modulus. The sol-gel transition is modulated by various factors, including UV dosage and polymer concentration. In some cases, gelation occurs in the "dark" phase, when the UV light is turned off, leading us to propose a two step-mechanism, involving calcium ion formation and its diffusion.

# Scalable Formation of Environmentally Benign Nanoparticles for Use As Delivery Vehicles of Active Ingredients

Authors:

Conner, C., North Carolina State University

Velev, O. D., North Carolina State University

Our group has introduced a class of environmentally benign nanoparticles (EbNPs) for biotech applications (Nature Nanotechn. 10:817, 2015). These nanoparticles are made of biodegradable lignin and have low potential for nano-toxic impact. The lignin core nanoparticles are synthesized through flash nanoprecipitation, but until recently they were produced in mL-scale batches. We have developed a semi-continuous system featuring a recycle loop, making it possible to produce such nanoparticles in practical quantities for industrial applications. We investigate the role of each variable in our process to determine how to control the size of our EbNPs and the final concentration of the EbNP suspensions. By changing the concentration of lignin in solvent, we are able to efficiently control the final size of synthesized nanoparticles. Then, by altering the anti-solvent volume, we control the final NP concentration of the dispersion. With our semi-continuous system, we are now able to make liters of concentrated nanoparticle suspensions at a time. We discuss a few applications of our EbNPs, including their role as highly efficient antimicrobials and antifungals. EbNPs that serve as antimicrobials are created by infusing silver ions into the lignin core and are coated with a polyelectrolyte layer, enabling them to stick to bacteria. This formulation has the potential to decrease use of silver in consumer products and thus limits its eventual release into environment. We also investigate the application of the EbNPs on other biological surfaces, such as rose petals and human hair. We modify the lignin nanoparticles with a cationic biopolymer coating, chitosan, to make them adhere to these surfaces. The nanoparticles can be functionalized with active ingredients as needed for different applications.

**An Emulsifier for High Sensory Formulas Containing up to 60% of Oil**

ECCP
Engineering
Cosmetics
& Consumer
Products

Authors:

Singh, P., Cargill Beauty

Belhaj, N., Cargill Beauty

Koch, C., Cargill Beauty

Verdu, B., Cargill Beauty

Ollagnier, J. N., Cargill Beauty

To emulsify high oil content in a minimalist approach has always been the challenging part to develop rich and nourishing finish products considering the sensory aspect with non-greasy and non-tacky after feel. StarDesign Power has been developed by Cargill Beauty to answer this customer demand and to bring to the market a wide texture palette. StarDesign Power is a patented combination of two modified starches from sustainable waxy corn origin. It was developed and optimized using a mixture experimental design of 70 trials. The objective of this study was to investigate the emulsification capacity of StarDesign Power and the effect of oil content (5 to 60%) on size distribution and stability of oil-in-water emulsions. Emulsions were prepared with 3, 4, and 6% w/w of StarDesign Power with and without hydrocolloid. Physico-chemical characterization has been performed to evaluate the stability of the formulas: Size distribution, viscosity, rheology profile and instability index from the Lumizizer. It was found that whatever the amount of StarDesign Power (3, 4 or 6%) with and without hydrocolloid, we observed an increase of viscosity starting from 30% of oil. This viscosity improvement is correlated to a decrease of droplets size and polydispersity. On the other hand, we have investigated the rheological behavior and the elastic network of emulsions with StarDesign Power at 3, 4 and 6% for 0, 30 and 50% of oil. The ratio G'-G" was increased by 2 from 30% to 50% of oil. The increase of the elastic modulus G' is highlighting the capability of the emulsion to recover immediately after a deformation and consequently to have better long-term stability. StarDesign Power decreases the interfacial tension and creates a viscous and suspending structured network in the water phase avoiding droplets coalescence and their migration and improving therefore the emulsion stability.

# Design of Sustainable Color Cosmetic Emulsions with Biosurfactants and Particle Mixtures

Authors:

Drakontis, C., Manhattan College

Color cosmetics have been gaining a lot of attraction in the market over the past few years and there are many new product formats which are constantly coming in. However, these are all emulsions which are stabilized by non-natural surfactants. There is still a very strong consumer driver to have the same performance benefits, in terms of long lastingness, emulsion stability, rheology, sensory etc. but delivered through sustainable materials. In this study we developed completely sustainable emulsions using mixtures of biosurfactants and silica particles. These formulations have been developed into a range of product applications, such as lip products, face masks and primers. Surface tension measurements, bulk rheological measurements of G'-G", viscosity vs shear rate, and interfacial rheology measurements were carried out on the biosurfactant mixtures together with the particulates. Synergistic interactions were seen which facilitate the emulsion performance in terms of emulsion stability, rheology, sensory etc. All product variations showed the efficacy which is expected. Different combinations of biosurfactants and particles will still need to be carried out to further improve the sensory, texture, and performance of these systems. We acknowledge Evonik for making this project possible.

# Designing High Performance Color Cosmetics through Optimization of Powder Flow Characteristics

Authors:

Liu, X., MANHATTAN COLLEGE

Amin, S., Manhattan College

Powder cosmetics constitute a large class of products which are increasing in popularity with consumers. Some newer product formats such as cream to powder or powder to cream products are also making their entry into the cosmetic market. In spite of the importance of color cosmetics, there hasnâ€™t been a systematic understanding of how powder flow and powder particle characteristics impact the final performance of these products ,such as pay off and cake strength. By using a Powder Rheometer, powder flow is quantitatively tested through flowability, compressibility, and shear tests. In this study, the composition and concentration of a powder foundation formulation was varied for a foundation formulation containing: Talc, Mica, $TiO_2$, yellow, red, black iron oxide. Ten formulations are prepared. The experimental variables were the concentration of $TiO_2$, Red iron oxide, and Talc. The results highlight that the formulations which has higher compressibility has the better cake strength, and the formulations which requires lower total energy has better pay off. Future studies will include exploration of the impact of addition of flow additives and variation in particle size and particle shape.
We acknowledge the contribution of Freeman Technology for making this project possible.

# Enzymatic Catalysis for Specialty Esters

ECCP
Engineering
Cosmetics
& Consumer
Products

Authors:

Burton, R., Lipid.Io

In commercial ester processing, the ability to process a wide range of feedstock grades provides favorable economics. Often cost-effective feedstocks contain high levels of free fatty acids (FFA) or other contaminants. Traditional chemical means of processing require sulfuric acid, high proportions of alcohol to fatty acid, and the infrastructure rectify sidestreams from such process. It is now proven that the immobilized and liquid phase enzymes can be utilized for the production of various food grade esters such as squalene. This paper will evaluate the range of the cosmetic processes that could be improved using enzymatic production.

# Formulation Design for a Completely Natural Shampoo

Authors:

Xu, L., Manhattan College

Amin, S., Manhattan College

The future of personal care and detergent products is the elimination of non-biodegradable, environmentally toxic surfactants. However, prior to that, an understanding of how performance criteria such as rheology is effected by substitution of traditional surfactants with biosurfactants is required. In this study, the effect of rhamnolipids (RLs) biosurfactant on the rheological response of traditional surfactants such as sodium laureth sulfate (SLES) and cocamidopropyl betaine (CAPB) is explored. In addition, sophorolipids (SLs) were introduced along-side with rhamnolipids to replace SLES/CAPB and form a completely natural shampoo. By utilizing mechanical rheometry and optical microrheometry, this study provides unique insights into microstructure-rheology linkages in complex biosurfactant-surfactant mixtures. Macro/microrheometry indicates biosurfactants/surfactants and mono/di-rhamnolipid ratio plays a critical role in the rheological response. Size data extracted from high frequency rheological responses indicates RLs concentration, electrostatic screening and pH has a significant impact on micelles structure. Mixture viscosity can be built up to a certain extent by adding salt and adjusting pH. A completely natural shampoo can be achieved by addition of hyaluronic acid (HA) to the RLs/SLs biosurfactants system. Effect of HA on the RLs/SLs system's rheology and microstructure will be further explored. We acknowledge LS Instruments, Evonik, and Lubrizol for making this project possible.

# Microstructure Design of Ctac:FA and Btac:FA Lamellar Gels for Optimized Rheological Performance Utilizing Automated Formulation Platform

Authors:

Davies, A., Manhattan College

Amin, S., Manhattan College

The cosmetic industry is very fast growing and mass customization has become an increasingly discussed topic within the cosmetic industry. As demand grows for more customized products and as consumers become increasingly aware of the ingredients that go into their various day-to-day products, the need for more flexible processing and manufacturing methods has increased significantly. This brings us to the utilization of automation across the cosmetic industry.

The main objective of this paper is to optimize hair conditioner performance through variation of composition utilizing automated cosmetic formulation platform and advanced characterization techniques. In this study, eight samples with varying quantities of Cetrimonium Chloride (CTAC) or Behentrimonium Chloride (BTAC) with Fatty Alcohol were made with the Chemspeed Flex Formax and through mechanical rheometry, rheological properties were studied. Controlled breakdown of the formulations (dilution) and its effect on wet combing were also carried out.

Yield Stress was found to have significant effect on dilution viscosity breakdown which also has a subsequent effect on wet combing test. This implies that overall performance of the hair conditioner can be engineered through optimization of the microstructure. Effect of substituting CTAC/BTAC with natural cationic surfactants will be further explored. We acknowledge Chemspeed, Dia-Stron and Evonik for making this possible.

# Nsoft: A NIST-Industry Consortium Focused on Neutron Measurements

Authors:

Weigandt, K., National Institute of Standards and Technology, NIST Center for Neutron Research

The nSoft consortium at the National Institute for Standards and Technology is a partnership between NIST and industrial consortium members. The consortium has several goals including the development of industrially relevant neutron measurement science, transfer of neutron scattering expertise to industry, and fostering collaborations between industrial partners and NIST. On this poster we will focus on sharing some of the measurement techniques we have developed or are continuing to develop within the consortium. Examples include: using deuterated vapor or solvents to resolve structures in films, use of automation and machine learning as tools for neutron scattering, measuring boron containing materials penetration into synthetic skin, and development of various rheo-scattering techniques.

Materials under flow are commonly observed in nature and industry, and it is therefore important to understand how deformation affects the structure of such materials. NIST has long history of developing rheo-scattering tools for neutron scattering experiments that are used to probe the structure of complex fluids under shear. These tools include rheometers, shear cells and other flow cells that have been developed or modified specifically to enable structural characterization during material deformation. RheoSANS has been used to elucidate the origin of strain hardening in certain biopolymer gels, measure shear induced structure formation in a range of micellar systems and colloidal suspensions, measure the evolving structure of polymers undergoing gelation, and more generally to explore structure-mechanical property relationships in a wide range of soft materials. Recently we have been focused on developing slit and capillary rheometer for high shear rate rheoSANS studies. In our newly developed instruments we will be able to probe shear rates in excess of $10^6$

# Profour: Taking Antioxidants to the Next Level

Authors:

Jordan, C. T., Bluegrass Advanced Materials, LLC

Guilfoil, E., Bluegrass Advanced Materials, LLC

Howerton, B., Bluegrass Advanced Materials, LLC

Dziubla, T., University of Kentucky

Hilt, J. Z., University of Kentucky

Wyse, J., Bluegrass Advanced Materials, LLC

Antioxidants have become a vital ingredient in personal care products due to their great anti-inflammatory and anti-aging properties; however, they are challenging to formulate, resulting in inactive consumer products with limited antioxidant benefits, or products that cause irritation by administering too high of concentrations to the skin over short periods of time. Four unresolved challenges with formulating with antioxidants are the instability of the ingredients, the limited window of activity, the lack of control in the amount delivered, and the low loading capacities of current delivery strategies. With these challenges in mind, BGA Skincare built the PRO*four* platform. PRO*four* is a novel delivery system that uniquely solves the problems associated with formulating any hydroxyl-containing active into skincare products. With a patented poly(beta-amino ester) technology, BGA Skincare can *protect* the activity of the ingredient by utilizing Michael addition chemistry and *prolong* its availability by delivering the ingredients for extended incremental time periods. The technology can also be *programmed* using tunable starting material parameters to achieve release profiles from 3-24 hours, and ultimately *provide* superior active loading capacities by incorporating the ingredient into the backbone of the polymer to achieve up to 5x greater loading capacities than other strategies currently on the market. PRO*four* will allow skincare products to fully capitalize on the known benefits of antioxidants, providing the full potential of the active to the consumer.

# AUTHOR INDEX